CHEMISTRY

CRITIQUE OF MODERN UNIVERSITY CHEMISTRY: RE-DEFINING CHEMISTRY

BY

M.F. ONUCHUKWU

To My Creator,
I give all reverence for the
understanding given to me to
write this great book.

Preface

The reader must do well to keep an open mind when reading this book because he shall be introduced to a new Chemistry altogether.

The reader must also endeavor to master the definitions given by the author, because it shall help in his proper understanding of the subject of Chemistry.

Finally, the reader must do well to read the author's book called the *"Absurdity of Numbers"*.

Table Of Content

Censuring Modern Chemistry – Chapter 1

The error of modern Chemistry lies

in the fact that it doesn't understand

the nature of Substance. In the

author's previous book, "*The*

Absurdity Of Numbers", the author

clearly reiterated that Substance is

the only thing that exists and that can

exist. It's in the nature of substance to exist. Nothing else in existence has that nature but substance.

According to contemporary chemists, modern Chemistry is defined as the study of matter and the changes it undergoes. Matter is defined as anything that occupies space and has mass. Also, according to them, substance is defined as a form of matter that has a definite (constant) composition and distinct properties. Before we go further with the other definitions that form the basis of modern university Chemistry, let's pause for a second and try to understand why the above definitions contain absurdities which make them untrue. First, from the definition of matter, the reader can safely come to an understanding on why substance can't be said to be a form of matter. According to the definition, matter occupies space.

Space in the sciences (including Chemistry) is the whole "area" in which all things exist and move. From the definition of space, it's clear that it doesn't involve infinity because of the fact that it uses the word "area". Area involves measurement and the author has proven clearly in the *"The Absurdity of Numbers"* that substance cannot be measured or quantified or counted, then it's safe for one to also conclude that area, space and matter consequently don't fall within the sphere of "existing". They don't fall within the context of Substance. The same goes for "mass" used in the definition of matter. From the explanation of the definitions of space and matter, the reader can clearly understand why it's safe to assert that the foundations of modern Chemistry were built on errors. The definition of substance according to modern chemists is based on the definition of matter which makes it very erroneous. They assert that Substance is a form of matter, forgetting that Substance is infinite and as such can't be measured or quantified or counted. It's also very absurd for the modern chemist to assert that substance involves being "definite". Substance can never involve being particular. Only the modes of substance are definite,

finite and temporal from a modal perspective; despite their essences involving infinity and eternity in reality. Also Substance cannot be said to involve constancy because involves duration and substance involves only eternity or existing. Also don't forget that Substance also has its states from which one can safely conclude that the modes of substance cannot be said to be in "constancy" too since they're always being modified in the different states in which they exist in substance. Composition can't also be said to be involved with substance because asserting so will mean that Substance can be divided into parts, which in reality is absurd. Substance can't be said to be composed of this and that because doing so will definitely mean that substance can be measured or quantified or counted, which is very impossible in reality. It's also quite false to assert that substance has distinct properties because it will mean that you can measure or count or quantify the essences of substance. The attributes of substance are always infinite. On the basis of the falsities described above, the modern chemist further extended the errors in his understanding to the definition of "mixture". According to him, mixture is a combination of two or more substances in which the substances retain their distinct identity. The problem with this definition of mixture lies in the fact that it asserts that there can be two or more substances. It's impossible to have two or more substances because that will make substance finite, which is an absurdity. This alone makes the definition of mixture very wrong. It's also impossible for Substance to "retain its distinct identity" because that will mean that substance has only one attribute, which isn't true. Modern Chemistry also asserts that mixtures can be homogeneous or heterogeneous. This assertion is false because substance can't be said to homogeneous because it will mean that substance can be measured or quantified or counted, which is absurd. Also, substance can't be heterogeneous because it will mean that substance is finite, which is also impossible. According to contemporary chemists, an element is a substance that can't be separated into simpler substances by chemical means. The falsity of the definition lies in the fact that substance can't be separated by any means whatsoever. Therefore, Substance can't be separated by chemical means. In modern Chemistry, a compound is defined as a substance composed of atoms of two or more elements chemically united in fixed proportions. Substance can't be composed of anything since it's only substance that exists. To assert that substance is composed of anything is to assert that substance is finite, which is a false statement. Substance is the only thing that exists. Nothing else exists but substance. Modern Chemistry also claims that "matter" exists in three states. This is very wrong because according to their erroneous definition of matter, it's impossible to go ahead and assert that it has states. The different states of substance are its modes. They don't lose the essence of substance despite being modified forms of substance or particular types or forms of substance. They only exist in substance. The contemporary chemist makes another fatal error in his definition of solids, liquids and gases. He tries to define solids, liquids and gases from the lens of volume and shape. He forgets that applying volume to substance is absurd because substance can't be measured or quantified or counted. He also tries to apply shape to substance forgetting that substance has infinite attributes of extension and as a result infinite shapes or forms (not modes) or outlines. According to modern Chemistry, solids can be converted to liquids and gases; but it doesn't take into consideration that there are states of substance (modes) which exhibit different characteristics. It doesn't take into consideration that solids, liquids and gases are particular types or forms of substance and as a result can necessarily be modified or changed into other forms of substance. It doesn't take into view the fact that the attributes of substance can be modified in a chain series of causes and effect. For example, if one took extension as an attribute of substance, he will soon discover that under that attribute, bodies exist and that their speeds can be modified and changed by other bodies in existence to infinity. The example cited above explains in reality the concept of solids, liquids and gases. A body is termed to be hard and solid because because the contact surface among its parts is large; it is difficult to change the position of the parts of the body; and the form of the body as well. A mode of substance is said to be soft because the contact surface among its parts is small and the tendency for its parts to change position becomes higher. Also, a mode of substance is said to be fluid or gas because its parts move about freely among one another. When solids come into contact with other bodies that are said to be "hot", they are modified; their parts move faster and they have less contact among themselves; the mutual relations between the parts become weakened and they change type or form (not shape). The same goes for soft bodies that come into contact with other hot bodies. They are modified, their parts move faster until there's no contact among the parts; and the mutual communications between them

becomes weaker until the body

changes type or form into fluids,

liquids or gases. If one asserted that

Chemistry is the study of how and

why bodies change form (modes)

when they come into contact with

other bodies, then one shall be very

close to the definition of Chemistry

in reality. Bodies can only be

identified by their properties and not

their composition. The composition

of all bodies remain the same because they are modes of substance (that is, they have in them the infinite essence of substance). The contemporary chemist asserts that "physical property" can be measured and observed without changing the composition or identity of a substance. This assertion is wrong because neither substance nor its modes (in this case "bodies") can be measured and observed because it's impossible to measure or observe what's infinite. Substance has infinite attributes which is what the human intellect perceives the essence of substance to be. In the case of bodies, the human intellect perceives the infinite essence of substance to be extension. It's also a farce to assert that substance can be identified. The identity of substance isn't narrowed down to a set of characteristics. The identity of substance always involves infinite characteristics (its essences) and they can never be changed. The mistake of modern Chemistry comes from the fact that it refuses to understand the differences between the essence of bodies and the properties of bodies. According to the contemporary chemists, substance can undergo a chemical change. However, it's impossible for substance to change because it will mean that there's another substance in existence which is an absurdity. The modern chemists also asserts that the properties of "matter" can depend on how much matter is being considered or how much matter isn't being considered. The key question at this point is this, "is it possible to measure substance?" The answer to the question is, "it's impossible to measure substance." The error of the modern chemist lies in the fact that he assumes that substance can be measured ot quantified or counted. This is a great anomaly in the understanding of substance and its modes. The chemist's poor understanding of bodies is shown in her definition of force as a quantity that causes an object to change its course of motion (either in direction or speed). The absurdity of the definition of force lies in the fact that it tries to measure or quantify or count the mutual relations between bodies. It tries to quantify how quick or how slow bodies are, which in reality is absurd. In other words, it's impossible to measure or quantify or count the infinite attribute of substance known as "extension". Bodies change how quick or slow they move or the direction of their motion as a result of the mutual relations they have with other bodies. Bodies only affect other bodies. However, bodies can be affected by a body in such a way that the form or characteristics of the bodies are moved in different forms by one and same body in accordance with its nature. In a similar manner, a body may be moved by different bodies in different forms or characteristics according to the nature of the bodies moving it. Modern Chemistry doesn't consider the modes of substance to be substance. This is an error on its own. All bodies are substance because they possess the infinite essence of substance which is in this case "extension". And as such, they can't be measured or quantified or counted. Neither can they be subjected to time because the attributes of substance are always infinite and "eternal". The modern chemist forgets that when bodies are analyzed from the viewpoint of eternity, it's safe to conclude that all bodies are eternally moving or eternally at rest. Another anomaly in the foundations of modern Chemistry, comes from the fact that the chemist considers energy to be the capacity to do work. He defines work as being done when an object is moved from one place to another in the presence of some external force. The absurdity of these definitions of energy and work lies in the fact that it's impossible to measure or quantify or count the motion of bodies. It refuses to take cognition of the fact that its in the character of bodies to be either at rest or in motion. From the previous definition of element, we discovered that it is absurd and fatally wrong to assert that it exists in reality. The same goes for the atom, which according to modern Chemistry is very much defined as the basic unit of an element which can enter into chemical combination. The definition of an atom shows that the modern chemist fundamentally misunderstood the nature of substance. He thought that substance can be divided into parts without considering the fact that substance is infinite. Atoms can't be said to be the building blocks of substance either because it will mean that substance is finite which in reality is an absurdity. From the infinite nature of substance, it's also safe to assert that the extent (if you may excuse the reader, in layman's terminology, the "size") of substance is also infinite as well. Thus, atoms or sub-atoms or whatever may be the result of the erroneous thinking that substance and its modes (in this case "bodies") can be divided and further divided to oblivion aren't the smallest of bodies in existence. You can't measure or quantify or count the extent of substance and its modes because they are infinite and eternal. The error in Chemistry like in the other sciences lies in the fact that they seek to divide substance which can't be divided in the reality. The conjuring up of "atoms" by modern Chemistry stems from the misunderstanding of substance by contemporary chemist. The modern chemist must come to the realization of the fact that Chemistry and other physical sciences involves only bodies and how quick or slow they are. The absurdities which beguile modern Chemistry is seen in the assigning of numbers to the so called "elements". It's also seen in the assigning of mass numbers to the so called "neutrons" and "protons". According to modern Chemistry, mass is the quantity of matter contained in a sample of substance.

The anomaly in that definition of mass lies in the fact that it's impossible to measure or quantify or count substance and the modes which exist in substance because substance is infinite. The contemporary chemist asserts that elements can be categorized as those that are metals, metalloids and non-metals. From the definition of elements, the chemist asserts that substance can be divided into simpler substances. This assertion is in itself an absurdity. The reader must recall that the study of bodies doesn't involve the nature of substances but the motion or rest of the said bodies and how quick or slow they are. Substance isn't the same as its modes even though its modes have its essence. On this basis, the chemist made a fatal error by categorizing substance as metals, non-metals and metalloids. It must be very clear to the reader that metals, non-metals and metalloids are all forms of substance and not substance as posited by chemists. The contemporary chemist took a step further in their fallacies by asserting that substance has definite chemical and physical properties as described in the periodic table. This is also an anomaly too because it's impossible for substance to have a chemical and physical properties for it will mean that substance was finite. All the "properties" of substance can be said to be infinite, if you considered the "properties" of substance to be its essence. Modern Chemistry makes another fatal error in its definition of molecule. It asserts that molecule is an aggregate of two or more atoms in a definite arrangement held together by chemical forces which may contain atoms of the same element or atoms of two or more elements joined in a fixed ratio. The definition of molecule is very preposterous because it first asserts that substance is made up of atoms. This is impossible because if anything exists apart from substance it will mean that substance is finite which is very absurd. Secondly, the definition of molecule asserts that substance can be joined in fixed ratio. This assertion means that there exists other substance with a similar nature like substance. This assertion is false because if it was true, it will mean that substance is finite and temporal which is an absurdity. The modern chemist made another grave mistake in his definition of chemical formulas as what combines symbols for the constituent elements with whole number subscripts to show the type and number of atoms contained in the smallest unit of a compound.

The error of employing chemical formulas in Chemistry stems from the definition of an element which describes it as a substance. From the realistic viewpoint, it's impossible for substance to be multiple because it will make substance finite and substance can't be finite but infinite. The employment of molecular formula by the contemporary chemist also shows his lack of understanding of substance and its modes. By assigning specific numbers to the atom (which he erroneously claims is the basic unit of an element), the modern chemist shows that he doesn't understand that substance can't be measured or quantified or counted. The invocation of chemical compounds into modern Chemistry depicts exactly what's wrong with the subject matter. The chemist defined chemical compounds as either molecular compounds (in which the smallest units are discrete, individual molecules) or ionic compounds (in which positive and negative ions are held together by mutual attraction). Ionic compounds are made up of cations and anions, formed when atoms lose and gain electrons respectively. The reader can succinctly see for himself the fatal error made by the modern chemist in assuming that substance can be split or divided into atoms, molecules or ions. He doesn't understand that substance can't be split or divided because substance is infinite. More mistakes are made by the modern chemist when he suggested that substance can undergo change that are termed chemical changes. And that these changes can be represented by chemical equations. If substance can undergo change, it will mean that substance can stop existing, which is very absurd. Also the reader, will agree with the author at this point, that the contemporary chemist's inability to differentiate between the substance and its modes has led to his believe that substance can undergo change. The author desires to mention this again, that Chemistry ought to involve bodies as well as how and why these bodies affect one another when they come into contact with one another. The efforts by the contemporary chemist to make Chemistry a science that involves the division of substance is very laughable to say but the least. It's sad to see the modern chemist make efforts to balance the so called atoms of the "divided substance" in the chemical equations he employs based on the erroneous laws of the conservation of mass. The author has proven why the conjuration of mass by the modern chemist is indeed a farce. It is a farce because its definition asserts that substance can be measured or quantified or counted which is an absurdity in reality. The contemporary chemist takes his misunderstanding of substance up another level when he assigns atomic mass (which he defines as the average of the naturally occurring isotope distribution of that element) to substance. He doesn't understand that substance can't be weighed or measured or quantified or counted. According to modern Chemistry, atom is the basic unit of substance (what the chemists call "element"). Also according to modern Chemistry, Avogadro's number is the number of atoms in an element. The key question at this point is this, "isn't it erroneous for Avogadro to have measured or quantified or counted substance?". The answer to the question is that it's very erroneous for anyone to assign values to substance because substance can't be measured or quantified or counted. Substance is infinite and eternal. However, we are confronted with a major fallacy in Chemistry called stoichiometry. Modern chemists define stoichiometry as the quantitative study of products and reactants in a chemical reaction. The problem with this definition lies in the fact that stoichiometry wants to "quantify" substance and its modes. In reality, it's absurd to want to "quantify" bodies. One can't "quantify" bodies, one can only observe the characteristics of bodies. Thus, if the contemporary chemist had understood substance and its modes, his objective will be to observe the changes in form that take place when the bodies affect one another and not try to quantify the bodies and the changes (which possess the infinite

essences of substance because they

are still substance in different forms)

they undergo.

Disparaging Units

of Measurement –

Chapter 2

Modern Chemistry adopted the

highest of fallacies and made it a

norm in the teaching of its subject

matter when it decided to measure or

quantify or count the modes of

substance. The same error was made

in the mathematics of numbers as

indicated in the author's book called

"The Absurdity of Numbers". It's

impossible to measure or quantify or

count substance and its modes.

Therefore, all the systems of

measurement adopted by Chemistry are done in error. The ancients were very interested in representing the modes of substance with what we can call "*characters*". The characters were generally pictorial figures drawn to represent these modes of substance. The ancients never measured or quantified or counted substance. Rather, they took great interest in representing the infinite attributes of "extension (or movement) and thought with these characters. The ancients perfectly understood that they can represent the modes of extension with characters that helped them remember exactly how and why that particular mode came into being. For example, the human foot was used by the ancients to represent the mode of extension (a particular type or form of the attribute of extension). For the ancients, if one drew ten feet on a wall or parchment paper, it will mean that "extension" wise, that he expected whomsoever will look at the picture to take ten equal foot strides and not that one was to measure or quantify or count the mode of extension which in this case is the earth. The reader must understand clearly what has been asserted so far. Also what the contemporary chemist failed to understand was that the so called "spheres of dimensional existence" are infinite. That is to say that it's impossible to measure or quantify or count any mode of extension, because they are infinite and eternal. The error made by the chemist was to assume that the distance of the earth can be measured in reality. This is a falsity and as such shouldn't exist. The platinum and iridium bars used as a standard for the "meter" is used as a result of a fatal error in understanding by the modern chemist. The platinum and iridium bars are modes of substance (in this case "bodies") and as a result they can't be measured or quantified or counted. They can only be distinguished from one another in respect of their motion and rest; quickness and slowness. The use of weight in measuring or quantifying the so called "effect of gravity" on bodies is laughable too. It stems from the same type of error that

occurs in the use of distance to

measure the "dimensions" of bodies.

It's impossible to measure or

quantify or count the effect of one

mode of substance upon another

another mode of substance. The

reader must also note that quick

bodies at rest tend to fall more

slowly when thrown down from an

elevated position than slow bodies at

rest when they are thrown down

from the same level of elevation.

Also moving quick bodies always

fall more quickly when dropped

from an elevated position than

moving slow bodies that are dropped

from the same level of elevation.

You only observe the effect and if

you choose you can represent the

modes of substance with

"characters" that show how inter-

related they are to one another. From

the first example cited in this

chapter, the drawings of the human feet will bring to the mind of the reader, the inter-relatedness between the human feet and earth. Volume is also erroneously employed by the chemist in measuring or quantifying bodies. The contemporary chemist defines volume as the amount of space that a substance fills. The definition of volume is an absurdity itself. If substance is infinite, how then can one posit that it's divisible, let alone capable of filling up the so called "space". Don't forget that the definition of "space" given in Chapter 1 was pointed out by the author to be an absurdity. The key question at this particular junction is this, why does the chemist insist on using volume to quantify substance and its modes (in this case "bodies")?. Let's now proceed to another fatal fallacy upheld by modern Chemistry. It's the fallacy of heat and temperature. For the umpteenth time, the reader must understand that hotness and coldness are the characteristics of bodies and can't be measured or quantified. To attempt measuring or quantifying the characteristics of the modes of substance is to attempt to measure or quantify the infinite attributes of substance. It's an impossible task. However, the reader at this point must understand that bodies are deemed hot or cold based on how quick or slow the oscillation of their parts are from their original positions. Bodies whose parts oscillate more quickly from their original positions are hotter than bodies whose parts oscillate less quickly from their original positions. It's an absurdity to try to measure or quantify how much displacement these parts make from their original positions. In this chapter, the author has proven that to measure or

quantify or count any mode of

extension is an absurdity in itself.

Disapproving Chemical Reactions – Chapter 3

In Chapter 1, the author extensively

proved that the definition of

molecules reeked with errors and absurdities. The reader shall agree with the author, that if the definition given for molecules was wrong, it will mean that any definition were it was used in will also be wrong. According to the modern chemist, chemical changes are the changes in the molecular composition of substance or of some substances in a mixture. If substance was to change its nature, it will mean that it'll stop existing, which in reality is impossible. Thus, it's imperative that the reader understand that the definition of chemical change is an absurdity itself. Also, the definition of chemical changes suggests that there can be "some substance". It's also farcical to remotely suggest that there can exist other substance besides substance. For this will mean that substance is finite, which is an absurdity. The nonsense of modern

Chemistry lies in the fact that it remotely suggests that: (1) there are more than one substance and (2) that substances can react. First, it's impossible for more than one substance to exist because that will mean that substance is finite which is an absurdity. Secondly, it's impossible for the self-caused substance to react with itself because it isn't in its nature to do so. Substance can only exist. However, the modes of substance affect one

another. And in the case of modern

Chemistry, only bodies affect one

another and in doing so change their

forms or type or characteristics (that

is they are moved in different

modes). For example when electric

current (a mode of substance) is

passed through a very dilute solution

of sulfuric acid in water, oxygen and

hydrogen gases are formed. What

the reader must observe about the

example above is that water (also a

mode of substance) changes its form

when affected by electric current and

sulfuric acid (another mode of

substance). That is to say that the

constituent parts of water (a body)

move more freely and quickly

among themselves when affected by

sulfuric acid and electric current. In

the same way, an iron nail (a body)

dropped in a solution of copper

sulfate (another body) causes the

constituent parts of the copper

sulfate to move more freely and

quickly changing form from blue to

colorless. Thus, the reader can

readily observe that the fluid copper

sulfate is affected by the iron nail. In

Chapter 1, the author clearly stated

that the existence of chemical

equations are an absurdity. The

contemporary chemist never

understood that it isn't in the nature

of substance to react but to exist. He

didn't also understand that substance can't be produced by anything but is self-caused. He also didn't understand that substance can't be measured or quantified or counted. His lack of understanding of the nature of substance made him to posit an "equation" that will explain the so called reaction between two or more substances. In reality this futile attempt to put together two or more substances is an absurdity in itself.

For example, when the body called magnesium (hard) comes into contact with another body called oxygen (fluid), the oxygen affects the magnesium in such a way that it causes the parts of the magnesium to move more freely and quickly among the already free moving parts of oxygen to form a different body called magnesium oxide. The reader ought to realize at this point that modern Chemistry erroneously believed that substance can be quantified and balanced in equations. It's okay to represent bodies with "characters" but it's very wrong to quantify these bodies like the contemporary chemist does. It's rather in the best interest of Chemistry and other sciences to tune their minds to understanding the motion and rest of bodies and their parts rather than going on a wild goose chase of measuring or quantifying them. The modern chemist takes his errors up another level, when he conjures the law of the conservation of matter into the world of Chemistry. He states in this particular law that matter can't be created or destroyed. The error in this law comes from the fact that the definition of matter is an absurdity. From the definition of matter, as anything that has mass and occupies space, the reader can understand why the invocation of matter into modern Chemistry is an absurdity. The definition of mass is an absurdity too. Mass is defined as a large amount of substance that doesn't have a definite shape or form. The problem with this definition of mass is that one can't measure or quantify substance. Also, shape or form doesn't fall within the premise of substance. Only, "existing" does. Thus, the law of the conservation of matter is nothing but a farce. It's nothing but complete nonsense. Only substance can't be created or destroyed. On the basis of the definitions of matter, the law of the conservation of matter shouldn't exist. Another fatal error employed in modern Chemistry is the law of definite proportions which states that substances react with one another in definite proportions and produce definite amounts of products. The error contained in this law lies in the fact that there only exists substance and nothing else. The existence of any other thing outside of substance will limit substance and make it finite. The error that stems from the misunderstanding the nature of substance has led the chemist to measure or quantify the modes of substance (in this case "bodies"). The author has also proven that substance doesn't react but exists.

The definitions of chemical reactions have been proven to be an absurdity. The author stated very clearly in the first Chapter of this book, that it was impossible to measure or quantify energy as a mode of extension. Hence, it's also a farce to admit that energy is released when "substances" reacts. The preceding statements are false because nothing else can exist but substance and substance doesn't react but exist.

Reproving The Kinetic Theory – Chapter 4

The contemporary chemist claims

that everything in the world is either

a solid, a liquid or a gas. The problem with this definition lies in the fact that it refuses to take into account the modes of other attributes of substance. It doesn't take into account the modes of thought and the other modes of extension. Thus, it's safe to assert that solid, liquid and gases are all states of substance. As earlier mentioned in Chapter 1, solid bodies have their parts making more contact with their positions firmly fixed and can't easily change their form. Soft bodies were also described as having their parts making less contact with one another and their positions lose and they can easily change their form. Fluid bodies (gases) were described as having their parts moving freely about one another. The author has reiterated for the umpteenth time that bodies are understood by their motion and rest. That's how quick or slow they are while in motion. In Chapter 1, the author also made it clear that the use of volume and shape in Chemistry was also absurd. That shape can only define bodies but can't be counted since there are infinite shapes (that's the mode of extension has the essence of the attributes of substance which are infinity and eternity) existing in substance. States of matter don't exist. Rather, the states of substance depend totally on the attributes of substance. Ice, water and steam are all states of substance (that is modes of substance). The author redefined temperature in Chapter 2, and laid emphasis on the fact that as a mode of extension, temperature can't be measured or quantified. Neither can pressure of gases be measured or quantified in reality because pressure is a mode of extension. If we are to describe pressure, we can assert that the parts which make up any gas move rapidly among one another and in the process of moving cause the parts of the body or bodies that they are in contact with to oscillate from their original positions. The modern chemist erroneously posits that a substance can diffuse. He forgets that diffusion isn't in the nature of substance. Only existence is. When the essence of substance is conceived by the human intellect to be extension (or movement), the only mode of substance to be considered is bodies. Only bodies whose parts move freely about one another can quickly move from one position to another while maintaining the mutual communication among themselves. For example, the parts of bromine will freely move about in a bottle filled with air because its parts move more quickly than that of air to produce a brown color. Thus, gases whose parts move faster than those of other gases can be said to be less dense than them. The invention of gas laws by the contemporary chemist shows that he doesn't understand the nature of substance. The Boyle's law, Charles' law and Dalton's Law of Partial Pressures all show that the chemist wrongly presumed that the modes of substance can be measured or quantified or counted. Also, Gay-Lussac's law is erroneous because it posits that gas has a volume. That is, that one can measure or quantify gas. This is an absurdity. If Gay-Lussac had posited that when gases come into contact with hot bodies, their parts will move quicker, he would be absolutely correct. His inability to simply explain gases in terms of the motion of their parts is the bane of his law. The author has clearly reiterated in the previous chapters that the definition of molecule is an absurdity because the definition of the atom is also an absurdity. The kinetic theory of gas is based on the definition of molecules and as such isn't right. The first assumption of the kinetic theory states that "the molecules of gas are relatively apart and have little attraction for one another". The problem with this first assumption of the kinetic theory is that it assumes that molecules are the same as the parts of gas. If the first law of the kinetic theory had stated that the parts of a gas move about freely among one another because their surfaces aren't in contact with one another, it would've been correct. The second assumption of the kinetic theory would've been correct if it stated that the parts of gases move about very quickly instead of the "molecules" of gases move about freely. The third assumption of the kinetic theory will also be correct if it assumed that the parts of gases move more quickly when they come into contact with hot bodies. The fourth assumption of the kinetic theory posits that molecules of gases are perfectly elastic. The parts of gases aren't elastic. Rather, when the free moving parts of a gas come into contact with one another, the parts moving quickly cause the other parts moving slowly to move quickly. The author also mentioned in Chapter 1 of this book, that the parts of liquids aren't in much contact with one another and that there was less difficulty for the parts to change their original positions. The contemporary chemist conjures up another fallacy when he asserts that viscosity can be measured. He forgets that all liquids (soft bodies) flow. That is, they move in a particular direction. Liquids will maintain a flow in a particular direction, so far as its parts maintain their mutual relations between themselves. From the definition of viscosity in modern Chemistry, it shall be very clear to the reader that the chemist doesn't understand the nature of substance and its modes. He defines the viscosity of a liquid as the length of time it takes for a given quantity of liquid to flow through a given tube. The errors contained in this definition of viscosity comes from the fact that it asserts that duration as a mode of eternity can be measured or quantified or counted. He also tried to measure or quantify liquid (which is a soft body and a mode of extension). If the chemist had asserted that the liquids whose parts have their surfaces less in contact with one another are less viscous than the liquids whose parts have their surfaces relatively more in contact with one another, he would've been absolutely correct.

Surface tension is another anomaly in modern Chemistry. It's very false to assert that one can remotely measure or quantify how bodies (modes of substance) affect other bodies. The how and why in the study of bodies lies solely in the degree (considering how quick and slow they are) of their motion and rest. The chemist defines surface tension as the force acting in the surface of liquids which tends to reduce the superficial surface to a minimum. For emphasis sake, the problem with the definition stems from the fact that it wants to measure how and why bodies affect themselves. Secondly, liquids have parts and its parts have surfaces.

Thus, a liquid shall remain a liquid so far the surfaces of its parts maintain their mutual communication with one another. Modern Chemistry invokes another nonsense when it outlined the kinetic theory of liquids which bleeds with unfathomable errors like the kinetic theory of gases. The kinetic theory of liquids make the same error like the kinetic theory of gases when it asserts in its first assumption that gases have "molecules" and that the molecules are relatively close together. If the contemporary chemist had asserted that the surfaces of the parts of gases are in little contact with one another with the tendency of the parts to change their relative positions more easily, he would have been correct. Also, the second assumption of the kinetic theory of liquids is very wrong. It erroneously asserts that the "molecules" of gases are moving. This assertion is a fatal error because it doesn't consider the fact that liquids have "parts" and not molecules because they are bodies. It also erroneously asserts that the parts of liquids are moving. The parts of liquids don't move about freely among one another. The third assumption which asserts that the movement of molecules increase with increase in temperature is also very wrong. If the modern chemist had asserted that the parts of liquids always oscillate more from their original positions when they come into contact with hot objects; and the contact surfaces they've among themselves shall continue to lessen with an increase in temperature until they become gases; he would've been totally correct without error. The fourth assumption that the collision of the molecules of liquids are perfectly elastic is also very wrong. It's impossible to measure the modes of extension. The fifth assumption which asserts that in liquids molecules exert some attraction for one another is very wrong. The parts of liquid have their surfaces not so much in contact with one another. To also assert that one can measure or quantify how and why the parts of a liquid affect one another is an absurdity. But the contemporary chemist erroneously conjures the so called Van Der Waals forces to erroneously explain why liquids have constant volumes and why molecules are held close together. The reader must not forget that in Chapter 1, the author asserted that the concept of volume was an absurdity in itself because it was impossible to measure or quantify the modes of substance (in this case the attributes of extension).

According to modern Chemistry, volume is defined as the quantity of 3-dimensional space enclosed by a closed surface. The problem with this definition is that it asserts that substance is 3-dimensional. This assertion is farcical because the attribute of extension can't be finite. It shall always be infinite. The definition of space is the boundless 3-dimensional extent in which objects and events have their relative position and direction. The error of this definition lies in the fact that it contradicts itself by asserting that space is both boundless and three-dimensional. That is to say that space is both finite and infinite. This is an absurdity in itself. Substance is the only thing that's infinite. All modes of extension are infinite. Thus, one can see that the definition of volume is convincingly wrong because it's a tautology. Solids as bodies ought to be distinguished from one another in terms of how quick or slow they are. Solids are hard because their parts are in contact with one another over large surfaces. That is to say that the contact surfaces of their parts among one another is large. The kinetic theory of solids is also very false. The first assumption of the kinetic theory of solids is very wrong because it asserts that in solids, the movement of the molecules is very slow. The parts of solids don't move at all. It's very difficult to change the position of the parts of a solid. The second assumption is also very meaningless because it asserts that Van Der Waals forces in solids are relatively very great. For the umpteenth time, It's impossible to measure or quantify how and why bodies affect one another. The third and final assumption of the kinetic theory of solids is fatally wrong because it asserts that the movement of molecules increase with an increase in temperature. If the modern chemist, had asserted that the parts of solids, oscillate more about their positions and that the contact surfaces between them diminishes when they come into contact with hot bodies, they would have been very correct. The concepts of evaporation, boiling, freezing, melting and sublimation can be

explained from the perspective of the

degree of motion and rest of the

parts of bodies.

www.ingramcontent.com/pod-product-compliance
Lightning Source LLC
Chambersburg PA
CBHW030948240526
45463CB00016B/2081